GUIDE TO THE BLACK COCKATOO

Covers Black Cockatoo history, feeding, species, habitat, nesting, & more…

Compiled and Edited by Roger Rainer

First Edition January 2017

Published by Wild Bird Press

© 2017 Roger Rainer

Rainer, Roger, 1950 –

ISBN 9781542990172

This book is copyright. Apart from any fair dealing for the purpose of private study, research, criticism or review, as permitted under the Copyright Act, no part may be reproduced by any process without written permission. Enquiries should be addressed to the Publishers.

All rights reserved.

Printed by Create Space, United States of America

CONTENTS

Contents

Introduction

Species of Black Cockatoos

The Size of Black Cockatoos

Black Cockatoo Beaks

Where do Black Cockatoos Lay Their Eggs?

Shyness in Black Cockatoos

Black Cockatoos in the Manly Warringah Area 1913

The Black Cockatoo Diet

Habitat of Black Cockatoos

Black Cockatoos in Western Australia 1928

Black Cockatoo Call

Black Cockatoos on the Darling 1877

Black Cockatoos Bred in Captivity 1945

Black Cockatoo Habits

White-Tailed Black Cockatoo

Comparison in Brief of the Different Species of Black Cockatoos

The Red-tailed Black Cockatoo (Calyptorhynchus banksi)

The Great Palm Black Cockatoo

The Funereal Black Cockatoo

Black Cockatoos in Tasmania

Leach's Black Cockatoo (Calyptorhyncus Leachii)

References

INTRODUCTION

There are many Australians who have never seen a black cockatoo, except perhaps at zoological parks. These fine birds belong mainly to the big timber and the solitary places of the country. They are shy birds and usually keep well away from people. Flying often in immense flocks, at a great height, they harmonise with the eerie weirdness of the gaunt, lofty gum trees which seem to reach out to them.

In the south-eastern corner of Australia the trees grow upwards of 200 feet or 300 feet, and the black cockatoos nest and rest in their topmost branches, appearing like imps of blackness to ground-tied humanity far beneath.

The favourite haunts of the yellow-tailed black cockatoos are the coastal districts and neighbouring mountain ranges, and the adjacent open forest lands from the neighbourhood of the Herbert River to north-eastern Queensland throughout eastern New South Wales into Victoria, and the eastern parts of South Australia and Kangaroo Island.

Usually they are seen in pairs or small flocks of four or five individuals.

Australians will perhaps be surprised to learn that there are no cockatoos in Europe, none in Asia except India, none in Africa north of the Tropic of Cancer, and only one in North America. There are various parrots found in other lands, but the only cockatoo known outside the Australian region is a species in the Philippines. We are fortunate here, and it is obviously our duty to protect unique bird life.

Like their snowy white cousins, the black cockatoos appear to live for 100 years and even longer. They favour smaller flocks, but may be seen occasionally in mobs of 50.

The black palm cockatoo of the Cape York jungles, is the giant of all black cockatoos.

All of the information contained in this book has been sourced from early Australian newspapers. The aim of this book is to take an historical look at black cockatoos, how they were viewed by people, their history, habits, species, diet, habitat, personality, identification and so much more. Some of the information may have been updated over time, but this publication will contain information on the various black cockatoo species, which is not well known.

I hope you enjoy reading and learning more about these beautiful birds.

Species of Black Cockatoos

There are five distinct species of black cockatoos in Australia. The funereal black cockatoo is one of the best known. Its ear coverts are of a dull yellow. This bird belongs to the genus Calyptorhynchus, which means a hidden bill, as the bill is often covered by the puffed-up feathers. All these birds always lay their eggs in the hollows of trees.

They live chiefly in the mountainous and heavily timbered country, and feed extensively upon the seeds in the pods, or nuts of the forest oak, eucalyptus, and other trees.

Their flight usually is slow and laboured, yet they can dive about between the trees in a most rapid manner.

Their call is slowly drawn out and mournful, and resembles "wy-ee-la". These birds cut large openings into the green eucalyptus trees with their powerful bills when hunting for big, fat, wood-boring grubs.

In the Dorrigo and Clarence River districts of New South Wales, there are trees eaten into by these birds for nearly ten inches when grub hunting. In such cases the ground beneath is strewn with large splinters. The birds are also fond of the flat grubs that live under the bark on recently rung, dead standing trees, and strip the bark off in great quantities when searching.

The five species of black cockatoos are:- the big palm cockatoo or the great palm cockatoo (Probosciger aterrimus) being the largest and the glossy or Leach black cockatoo (Calyptorhynchus lathami) being the smallest. The others are the red-tailed or Banks black cockatoo (Calyptorhynchus banksi), the yellow-tailed or funereal black cockatoo (Calyptorhynchus funereus) and Westralia's white-tailed or Baudin's black cockatoo (Calyptorhynchus baudini). The word Calyptorhynchus signifies a hidden bill, as it is often covered by the puffed-up feathers.

Except the palm cockatoo, which inhabits the Cape York Peninsula and also occurs in the Aru Islands and New Guinea, the others are found throughout Australia. They are somewhat nomadic, their movements being regulated by the food supply, the seeding of the different species of banksias, casuarinas, etc,.

The yellow-tailed or funereal black cockatoos and the red-tailed (Banksian) are the best known species, while the big palm cockatoo seems to be unknown outside the Cape York district. However, all the black cockatoos are to be met with in Queensland, with the exception of that strange West Australian, the white-tailed. They are picturesque birds, which claim the whole attention whenever they are seen. An experienced old bushman of the West, claimed that these birds were reliable bush barometers, forecasting storms when they came in after wattle grubs.

The common Funereal Cockatoo has a yellow-banded tail whereas the others have tails barred with red or white feathers.

None of the five are really black, black, though, but dark brown or sooty blue or both.

The white-tailed black cockatoo is the only one not found in Queensland. Black cockatoos are quite good talkers if kept as pets. The great palm cockatoo is the only black species with a real crest.

The Size of Black Cockatoos

Black cockatoos are larger birds than the white except for the little glossy black.

The black cockatoo is a magnificent bird. It measures about 30 inches, of which the tail is ten inches.

Black Cockatoo Beaks

Their beaks are specially adapted by nature for biting through wood and other hard substances to enable them to obtain the food upon which they mainly subsist.

The largest black cockatoos are found in New Guinea. Its most curious feature is its bill which enables it to live on the Kanary Nut, which is found on the highest trees. This nut is so hard that only a heavy hammer will break it. No other bird can eat it. The cockatoo cuts a notch in the nut by a sawing motion of his lower bill. It then breaks off a piece with a sharp nip. It's very long and very sharp upper mandible is inserted and the kernel is eaten.

All black cockatoo species have extremely large beaks and jaw muscles so powerful that the birds can crack the triangular kanary nut which just about defies a table nut-cracker.

Besides nut cracking, the birds use their beaks for digging holes in the trunks and branches of trees to extract witchety grubs.

Their bills are exceptionally powerful, as one may judge by the manner in which they tear to pieces the bark and wood of infested trees, often to depths of six inches into green trees, seeking grubs, and the ease with which they crack hard seed-cones.

Where do Black Cockatoos Lay Their Eggs?

Black cockatoos choose a hole in a very tall gum, very often a dead and rotten tree, the eggs will be found about six feet down inside a hollow.

The males feed their sitting mates in the nesting season as lovingly as little yellow robins do.

Nesting time is in July, August and September.

The young ones are very difficult to rear, and very noisy.

(Above - A three week old baby black cockatoo.)

The first recorded egg of the great palm cockatoo (of North Queensland, New Guinea, and the Aru Islands) came from Papua. Some years later in 1897, a specimen was collected at Cape York – the first authenticated egg of the giant black cockatoo found in Australia. As a rule, one egg forms the clutch. The favourite nest site is a piped branch or a hole in a tree. The nesting season extends from about November until March, so that the birds rear their broods during the wet season.

Around 1880, nests of the Banksian black cockatoo, which ranges from Queensland to Victoria and South Australia, were found in Western Queensland, nesting in May and June. The eggs seen were the first authenticated sighting. The nests were located in tall gum trees on the Darling.

Nests of the red-tailed cockatoo were discovered by the Horn Scientific Expedition to Central Australia in 1894. Many pairs were found nesting on the Finke and other rivers.

They nest in holes in trees, usually at great heights from the ground, and one or two white eggs are laid for a sitting.

These birds are not abroad by day, but nesting and roosting in places, which may be located by looking for pellets of food remains, which the birds vomit out of the hollow onto the ground.

The most extraordinary thing about the hatching is, the young are looked after by the parents until nearly fledged.

Shyness in Black Cockatoos

Black cockatoos are excessively shy, and will only nest in dense, high forests.

Shy, retiring birds, they prefer the quiet bush recesses, away from the haunts of men. They are restless birds, and rarely remain long in any tree.

They are very shy birds when they have a nest, and it is not often that an opportunity is obtained of observing one going to or coming from the nesting hollow.

The great palm black cockatoo is a very shy bird and feeds chiefly upon the tender tops of palms, attacking the hearts.

Black Cockatoos in the Manly Warringah Area 1913

The very large yellow-tailed black cockatoo was very numerous in the Manly and Warringah area around 1913, during the winter and early summer time.

They would travel in flocks of 10 to 15, wide distances apart, keeping to the coast line and feeding from off the ground on a certain seed shed from one of the hardest nuts of a native plant.

The birds were rarely molested by residents, and fed close alongside residences at times.

There was always one of the birds on guard to give the alarm note on the approach of danger.

The Black Cockatoo Diet

Their principle diet consists of grubs, borers, etc., which are found in the trunks and limbs of trees, green and dead. These fine birds, however, are friends rather than enemies of the forester. If they sometimes damage trees they compensate with compound interest by the ceaseless war they make upon insect enemies of the forest timber.

Black cockatoos feed on gum nuts, but if forest food is short they will feed on whatever they can.

Yellow and red-tailed black cockatoos are very fond of the Cullagolla or swamp sandalwood seeds.

Black cockatoos feed on the seeds of the eucalypts, banksia, and casurinas, and on large white beetles and timber insects and larvae.

The black cockatoos are most useful birds, for they are nearly always waging war against the boring insects and pupe thereof. These destructive insects are a great menace to our forest trees, and the black cockatoo was created to keep them in check. The birds soon locate the insects under the sappy bark of the trees and branches, and with their strong bills extract the boring insects and their lava from their tunnel-like auger holes.

By eating the tree pests, black cockatoos promote the growth of trees and preserve the life of the timber tree.

One of the black cockatoos favourite foods is the kernel of the double-gee, which they obtain by cracking the seed with their beaks. This cracking can often be heard some distance away. With their powerful beaks they break open very hard nuts which by us can be broken only with a hammer. These nuts grow on bushes found on scrub plains.

Most of the year the birds in question live in the ranges among the big timber, where they nest in the hollow trees and rear their young. When the insects are in the egg stage, therefore, and the supply of food is scarce in the hills, the birds migrate south, the kidney-shaped seeds of the Hakea bushes are ripe in March. They will often congregate in small parties on the burnt scrub ground to feed on the roasted seeds of the hakea, which have cracked after the heat, thus affording a much easier extraction of the seeds.

Like all cockatoos their feet are used as hands, and one marvels at their expertness in picking out the seeds from casuarina cones, which humans find require a hard blow of a hammer to dislodge. They can take a single canary seed in their bill, crack it in perfect halves, and extract the kernel. We could not do it with the finest of instruments; yet they perform the task in a twinkling, rarely dropping a seed.

Black cockatoos have no trouble in cracking the green, unripened and tough pods of hakea. Hakea and banksia are principal items of diet of funereal black cockatoos, with casuarinas and wood grubs.

The great palm black cockatoo love to eat the kernel of the kauari nut, the shell of which is said to be harder than that of any other nut, and to protect a kernel of the most delicate flavour. The Kamari tree grow to a great height and bears a fleshy fruit which is enclosed in an extremely hard shell of glass-like smoothness of surface. Within this shell are from one to three kernels covered with a thin skin; when this is removed the nut falls into a number of irregular flakes of snowy whiteness and delicious taste. The fleshy part of the kauari fruit is eaten by many birds, but only the black cockatoo is able to get at the nut, which it does by the great strength of its immense, sharply-pointed and beoked beak.

Taking a nut end in its bill and keeping it firm by a pressure of the horny end of its tongue, the cockatoo cuts a notch across the shell by a sawing motion of the sharp edge of the lower part of the beak. This done, the bird takes hold of the nut with one foot while biting off a piece of a thick leaf; this it wraps around the nut to prevent the glassy shell from slipping, while it uses the upper under part to insert in the notch already made and wrench off a piece of the shell by a powerful nip.

Again taking the nut in its claw the bird inserts the very long and sharp point of its bill into the hole just made and picks out the kernel which is seized flake by flake by the horny end of the long and flexible tongue.

More time is required to tell about this nut cracking, than the bird takes to perform the operation, for the cockatoo is a very rapid feeder and will consume a great many nuts in an hour.

The black cockatoos will laboriously crush the large gum nuts in order to reach the small seeds inside.

Black cockatoos will also eat blue gum saplings.

Habitat of Black Cockatoos

Black cockatoos are mountain birds – that is to say they confine themselves almost entirely to the mountainous regions of the continent.

In Tasmania they are regarded as most reliable 'weather prophets'. When they come down from the mountains to the lower country in large numbers it is almost invariably a sign of stormy and rainy weather. Tasmanian bushmen accept their appearance in these circumstances as a sure indication of heavy rain.

The red-tailed black cockatoo also inhabits Central Australia as well as being a denizen of the West and North-West.

Black Cockatoos in Western Australia 1928

In 1928 black cockatoos were seen in swarms of hundreds in the country districts of West Australia. Towards dusk, when they are settling down to roost, they make a great clattering noise, and will squabble over coveted positions on the branches, where many an encounter takes place and beaks and claws were bought into play. They were a great menace to farmers during the sowing season, because being so numerous, they would soon strip large areas of newly-sown fields.

Black Cockatoo Call

The voices of black cockatoos are usually described as harsh, but to my ear the melancholy double-note call of a small company of them flying over at a distance is pleasant, strangely suggestive of loneliness.

They never screech like white cockatoos.

Black Cockatoos on the Darling 1877

In the year 1877, a young man, whilst stuck on a steamer on the Darling, near Jandra station, 12 miles below Bourke, was witness to a myriad of black cockatoos. They suddenly appeared one day with deafening screeches, and almost at once began nesting in a low box timber.

There were often two and 3 nests in one tree, some of them only about 8 or 10 feet from the ground. The steamer remained in this position for several months while the river was low.

Black Cockatoos Bred in Captivity 1945

In 1945, reports appeared regarding the first person ever to breed an Australian black cockatoo in captivity.

In the aviary at his Sydney home, Mr Halistrom had 22 black cockatoos, among them were the red-tailed, yellow-tailed and Great Palm.

Black cockatoos are one of the few species of birds where the female is more beautiful than the male.

Black Cockatoo Habits

In habits black cockatoos closely resemble one another, usually being observed in pairs or small flocks, and are often exceptionally shy and wary, flying shrieking away at the slightest approach of an intruder. Their food consists chiefly of the seeds of trees and the larvae of insects which infest our forest trees. They are very partial to the large white grubs, the larvae of the goat-moth (Eudoxyla eucalypti), which does considerable harm to acacias, casuarinas, banksias, and many hardwoods. These grubs are also a favourite delicacy of the aborigines, and many bushmen who have tried them declare that they are excellent eating when roasted. It is remarkable how much litter the birds make when feeding. The ground beneath a tree is literally covered with the refuse of cracked seed-cones, green leaves, twigs, blossoms, and small branches or pieces of bark and wood, telling at a glance what they have been feeding on.

The black cockatoo is typically seen alone, or at the most in twos and threes. The Banksian (Calzplorlynchus banski) is found in the mountainous country and dense forests of all the States, particularly Kimberley, the Territory and North Queensland; in fact it is a northern bird. The yellow-eared (Calzplorlynchus funereus) is in a habitat of the south only, especially is the Victorian Grampians, but they are very rare and shy.

The choice of habitat of the black cockatoo is dictated by the character of their food. There is a huge grub in the Kimerberley that lives in trees locally called by the Aborigines 'birdies'. They are very tasty and when cooked on an ash fire resemble custard. The black cockatoo has a taste for these grubs and trees in Kimberley are often denuded of their bark by these birds in their search for them.

White-Tailed Black Cockatoo

The White-Tailed Black Cockatoo, Calyptorhynchus baudini shares with its near relative, the red-tailed black cockatoo, or Karrack, the distinction of being the largest 'parrot' of our South-West. It is the common black cockatoo of the Perth district and of the coastal plain in general, for the red-tailed species is essentially an inland form, though it does at times visit the coast. I have seen an occasional bird near Perth, and have seen fairly large flocks of them near the Margaret River caves, where they formed most attractive pictures on the tall gums.

The bird, when laid out flat, measures over two feet in length, of which about 10 inches is accounted for by the tail. The general colour of the head and back is blackish-brown, but a closer inspection shows that each feather on the hind neck, shoulders, and back has a paler margin. The crown of the head, the crest, the sides of the face, and the throat, and the long flight feathers of the wing, are uniformly blackish, but the rest of the under-surface is similar to the back, through rather paler, each feather again having a light margin. There is a patch of pale straw-yellow feathers covering the ear, and the middle portion of the outer tail feathers are dull white, except the outer edges and the shafts, which are blackish.

In this species there is no distinction between the sexes, except that the full-grown male is rather larger than the female. The same difference also applies to the karrack, which, by the way, often exceeds the white-tailed black cockatoo in size by an inch or so.

Being a West Australian species, it is not surprising that the bird was not known to science until Lear figured it in 1832. Probable specimens were sent home soon after the foundation of the colony.

They associate in flocks when the breeding season is over, flying rather slowly with a flapping flight, and constantly uttering their mournful piping note, which resembles the word oo-lack, which is the aboriginal name for them in the extreme South-West and around Broomehill. They feed largely on the seeds of the red gum trees, to obtain which they bite and tear open with their powerful beaks the very hard nuts that contain them. At times a flock will settle in an orchard, and do great damage in a very short time to the buds and young shoots of the trees.

The breeding season about Broomehill seems to commence in August and continue until November. Two eggs appear to be the full clutch, and very frequently one egg is not hatched.

In 1929 a man who had discovered several nests stated that the second egg was laid about a fortnight after the first and was not sat on after the first chick was hatched. The nesting cavity is usually a considerable height from the ground but this man did not see one more than 12 feet from the ground that contained eggs.

The fledged young birds are fed by the parents for many weeks after leaving the nest.

In 1912, flocks of white-tailed black cockatoos would visit the wonderful jarrah and karri eucalyptus forests at the Bow and Frankland Rivers, in South Western Australia. When flying, their white tails were beautifully spread out. The bloodwood eucalyptus trees there bear quantities of hard oval pods, measuring about two inches long. These were emptied by the black cockatoos, who would sit in the trees for hours patiently holding them in their claws and picking out the small unripe seeds.

The white-tailed species is brownish black in main body colour. There is a broad white band across the tail and a large white cheek patch. This particular type is confined to the south-west corner, up until 1950 the present range of these birds was from the Murchison to east of

Esperance. The inland extension is as far as Wongan Hills, Kellerberrin and Norseman.

The bill of the white tailed species is also an admirable instrument for tearing the bark of the ring-barked gums in order to get at the insects and grubs underneath. It is astonishing to see how rapidly the bird will work and how quickly the pile of shreds grows in size at the foot of the tree.

Comparison in Brief of the Different Species of Black Cockatoos

Banksian

Male

The tail band rich vermillion.

Female

Tail band red passing into sulphur yellow on edges, freckled with black; under tail coverts freckled bars of yellow-red; head, neck, wing-coverts yellow tipped; narrow bars of yellow on under-surface.

Young Males

Young males somewhat similar.

Great-Billed

Male

Has shorter wings than banksian, and bill one-third longer; tail band scarlet.

Female

Differs from banksian only in tail band, those parts which in the banksian are pure scarlet in this bird are mingled yellow and scarlet.

Western

Male

Has a shorter and more gibbous bill, more rounded crest, and shorter tail than last bird.

Female

Tail band dull scarlet, crossed by many bars of black; under tail-coverts crossed by bars mingled yellow and scarlet; tip of feathers of head, cheek, crest, and wing-coverts spotted yellow; under part has numerous irregular yellow bars.

Leach's

The last species, and most swollen gibbous bill.

Male

Tail band broad; scarlet or vermillion in colour.

Females and Young Males

Have the scarlet bands of tail crossed by narrow bands of greenish black, yellow feathers sometimes on head and cheeks.

Both Sexes are Alike:-

Funereal

Largest of genus, but not so large as great palm microglossum.

Ear-coverts, wax-yellow; tail bands brim-stone yellow, thickly freckled with irregular zig-zag marking of brownish black; outer web of outside tail feathers brown.

Yellow-Eared

Vary in size and weight.

Ear-coverts bright-yellow; tail band light lemon-colour, in some specimens this is blotched with brown spots; the yellow goes quite across the tail and does not stop at the outside web as in the funereal.

Baudin's

Tail band white.

Microglossum

No tail band whatever, tail black; largest of all and longest crest.

The Red-tailed Black Cockatoo (Calyptorhynchus banksi)

When speaking of black cockatoos one must always bear in mind that there are two species of these birds in the South-West portion of this State – the red-tailed and the white-tailed varieties.

The former is found practically throughout Australia and, although missing from Tasmania, is present on King Island in Bass Strait. The latter is found only in Western Australia, and appears to have its northern boundary somewhere in the vicinity of the Murchison River. It is probably the better known of the local species, as it is not uncommon in the metropolitan area at certain seasons, when its piercing cries are easily recognised.

The distinguishing feature of most value is implied in the names of these birds, as a band of colour crosses the tails of both sexes. The red-tailed variety measures about 2 feet in length and the male has a beautifully, glossy black plumage, but the female is rather more brownish owning to the light edges of the feathers, and is characterised by a tinge of yellow on the sides of the head.

In the South-West the black cockatoo inhabits the vast red gum and jarrah forests, feeding upon the seeds which they bear, as well as the various wood-boring grubs which infest their trunks. Further north and inland the white gum timber forms their home, and in the extreme north they feed upon the dwarf eucalypts and a tree known locally as the chestnut, which bears large quantities of stony fruit with hard seeds somewhat resembling those of a date.

The insect-eating propensities of these birds often bring them into disfavour, for in searching for grubs in the tree trunks they often gash the limbs about rather seriously.

The cry of this species is a harsh grating note, which is uttered almost incessantly, whether its producer is flying or at rest.

This species was the first discovered and is the best known of our black cockatoos. It is named after the celebrated naturalist Sir Joseph Banks, who was the friend and companion of Captain Cook when voyaging among our Australian waters.

THE RED-TAILED BLACK COCKATOO

In 1880, it was observed that the Banksian species did not congregate in large flocks, and it was believed to be solely arboreal in its habits.

The female and young birds of both sexes differ considerably from the old male I the marking of their tail. The male has the entire plumage glossy greenish-black, with a broad band of rich deep vermillion across the middle of all but the two central tail-feathers, and the external web of the outer feather on each side; feet nearly brown; bill in young specimens greyish-white, in old black. The female has the general plumage glossy greenish black, each feather of the head, sides of tail, neck and wing coverts pale yellow; under surface crossed by narrow irregular bars of pale yellow, becoming fainter on the abdomen; under tail-coverts crossed by narrow freckled bars of yellowish red; tail banded with red, passing into sulphur-yellow on the inner margins of the feather, and interrupted by numerous narrow irregular bars and freckles of black. Some people think they are glossy greenish black others a dull brownish black.

The Great Palm Black Cockatoo

Our largest form is the Great Palm Black Cockatoo (Misroglossus aterrimus), of Cape York, Queensland, and New Guinea; and is a handsome bird. All this family live chiefly in the mountainous and heavily timbered country, and feed extensively upon the seeds in the pods, or nuts, of the eucalyptus, banksia, and forest oaks (Casuarina), etc, and in one of the latter trees the writer has often seen a flock of them feeding peacefully, and almost silently, until they fly off, when there is a perfect uproar. Their flight usually is slow and laboured, yet they can dive about between the trees in a most rapid way.

The great palm black cockatoo of Cape York and New Guinea renders a low, short whistle, and possesses an immense bill, and a long blackish crest. It lays a single white egg, seldom two, in the hollow of a tree, and in order to avoid the danger of the hollow getting flooded during heavy tropical rains, this bird actually splits pieces of wood into thin strips about three inches long and drops them down into the hollow for several inches. On this heap the egg is laid. Should water get into the hollow the pile of criss-cross strips of wood acts as a protecting platform, and prevents the water from reaching the egg. All the others species of black cockatoos lay from one to two white eggs, and very rarely three and nest in hollows of trees, the eggs being simply laid upon the decayed wood. They often enter a hollow high up and go down inside the trunk until they are almost level with the ground, and there deposit their eggs.

They are very shy birds, and feed chiefly upon the tender tops or hearts of the various palms. All this family live chiefly in mountainous, heavily-timbered country and feed extensively upon the seeds in the pods, or nuts, of the eucalyptus, banksia and forest oaks (Casuarina). They also eat into the trunks and branches of the trees, to feed on wood-boring grubs.

The Funereal Black Cockatoo

The Funereal black cockatoo is one of the best known; its ear coverts are of a dull yellow, and the tail feathers banded with yellow, and well freckled with curious undulating marking of blackish-brown, while some of the other species have red or scarlet tail feathers.

Sometimes before rain this bird flies very low down over the tree tops, repeatedly rendering its slowly drawn out and mournful call.

A particularly interesting point is that these birds when thus engaged in grub hunting, they cut and strip the wood out perpendicularly running with the grain.

On the top of Mount Woolooma, near Bell-trees, Upper Hunter River, New South Wales, you can find rings neatly and deeply cut into the bark, and almost encircling the trunks of several large stringy bark eucalyptus trees, the work of black cockatoos.

A peculiarity regarding these birds is that if one of a flock is killed or wounded the others will fly round and repeatedly return to the same tree.

The funereal black cockatoo is one of the best known. Its ear coverts are yellow, and the tail feathers banded with yellow, and flecked with curious brown marks, while some of the other species have red or scarlet tail feathers.

The aborigines know the bird in some districts, by the name of "Wy-ee-la" and "Curra-naa", which resembles its cry.

A flock often settles in a large tree, and nips off masses of leaves and twigs in a very short time.

Black Cockatoos in Tasmania

The black variety found in Tasmania is an insular form of the funereal black cockatoo (Calyptorhynchus funereus), found on the mainland, and at one time was classed as a different species.

During winter, it is not unusual to see small flocks flying over Hobart.

Leach's Black Cockatoo (Calyptorhyncus Leachii)

This species is almost as well known as the Banksian variety, and as far as Queensland is concerned it is the most commonly seen of all the varieties. It is a bird of the interior as well as of the coast, and is found anywhere where there is timber, from Carpentaria to Victoria. It is not a regular migrant, but, like all the parrots, it follows its food, and according to the plentifulness of this so is it to be found. The Casuarina (oaks) is almost essential to its existence, for that its food is the seed of these trees; but in this colony the bloodwoods and other eucalypts are quite as frequently visited by these birds for the purpose of eating the seed-vessels of those trees. This species is also a grub-eater.

The *C. Leachii* is the least species of all the black cockatoos, and independently of its smaller size may be distinguished from its congeners by the more swollen and gibbous (conoes protuberant) form of bill. It is less shy or distrusting than are the Banksian and Funereal species, but little stratagem being required to get within close quarters.

The old male has the entire plumage glossy greenish black, washed with brown on the head and neck, with a broad band of deep vermillion across the middle of all but the two centre tail-feathers, and the external web of the outer feather on each side; irides, very dark brown; orbits, mealy black in some, in others pinky; bill, dark horn-colour; feet, mealy black.

The females and young males differ in having the head and neck browner than in the adult male, and in having the scarlet band on the tail crossed by narrow bands of greenish-black. It is not unusual to find individuals of this species with yellow feathers on the cheeks and other parts of the head; this variation is subject to no law.

The End

REFERENCES

1945 'Breeds rare birds to save them from extinction', *The Sun (Sydney, NSW : 1910 - 1954)*, 13 May, p. 10. , viewed 07 Feb 2017, http://nla.gov.au/nla.news-article230442864

1932 'BIRD WEEK—SCIENCE CONGRESS—SPORT', *Sydney Mail (NSW : 1912 - 1938)*, 24 August, p. 1. , viewed 07 Feb 2017, http://nla.gov.au/nla.news-article166225195

1930 'The Naturalist', *Western Mail (Perth, WA : 1885 - 1954)*, 13 March, p. 43. , viewed 07 Feb 2017, http://nla.gov.au/nla.news-article38521719

1924 'A STUDY IN ORNITHOLOGY', *Countryman (Melbourne, Vic. : 1924 - 1929)*, 26 December, p. 13. , viewed 06 Feb 2017, http://nla.gov.au/nla.news-article223721551

1890 'The Naturalist', *Bathurst Free Press and Mining Journal (NSW : 1851 - 1904)*, 22 November, p. 5. , viewed 06 Feb 2017, http://nla.gov.au/nla.news-article65346462

1922 'Life and Lore of the Bush', *Sunday Times (Perth, WA : 1902 - 1954)*, 14 May, p. 14. , viewed 06 Feb 2017, http://nla.gov.au/nla.news-article58029688

1880 'Psittacl.', *The Brisbane Courier (Qld. : 1864 - 1933)*, 9 June, p. 5. , viewed 06 Feb 2017, http://nla.gov.au/nla.news-article890566

1939 'FUNEREAL YELLOW-TAILS AND GANG GANGS', *The Australasian (Melbourne, Vic. : 1864 - 1946)*, 25 March, p. 41. , viewed 06 Feb 2017, http://nla.gov.au/nla.news-article141815612

1950 'LORE LEGEND', *Western Mail (Perth, WA : 1885 - 1954)*, 26 January, p. 15. , viewed 06 Feb 2017, http://nla.gov.au/nla.news-article39100329

1880 'Psittaei.', *The Brisbane Courier (Qld. : 1864 - 1933)*, 29 May, p. 3. , viewed 06 Feb 2017, http://nla.gov.au/nla.news-article895372

1944 'The Black COCKATOO', *The Mercury (Hobart, Tas. : 1860 - 1954)*, 2 September, p. 9. , viewed 06 Feb 2017, http://nla.gov.au/nla.news-article26029435

1938 'THE BLACK COCKATOO', *The World's News (Sydney, NSW : 1901 - 1955)*, 11 May, p. 20. , viewed 06 Feb 2017, http://nla.gov.au/nla.news-article137001335

1933 'AUSTRALIAN BIRDS.', *The Sydney Morning Herald (NSW : 1842 - 1954)*, 1 July, p. 9. , viewed 03 Feb 2017, http://nla.gov.au/nla.news-article16987313

1952 'Juniors' Corner THE BLACK COCKATOO', *Queensland Times (Ipswich) (Qld. : 1909 - 1954)*, 15 March, p. 4. (Daily), viewed 03 Feb 2017, http://nla.gov.au/nla.news-article124587291

1933 'Life and Lore of the Bush', *Sunday Times (Perth, WA : 1902 - 1954)*, 19 March, p. 12. (Second Section), viewed 03 Feb 2017, http://nla.gov.au/nla.news-article58676522

1934 'Life and Lore of the Bush', *Sunday Times (Perth, WA : 1902 - 1954)*, 4 November, p. 11. (Second Section), viewed 03 Feb 2017, http://nla.gov.au/nla.news-article58730095

1884 'COMPARISON IN BRIEF OF THE DIFFERENT SPECIES OF BLACK COCKATOOS.', *Seymour Express and Goulburn Valley, Avenel, Graytown, Nagambie, Tallarook and Yea Advertiser (Vic. : 1882 - 1891; 1914 - 1918)*, 22 February, p. 4. , viewed 03 Feb 2017, http://nla.gov.au/nla.news-article165087210

1929 'The naturalist', *Western Mail (Perth, WA : 1885 - 1954)*, 2 May, p. 50. , viewed 02 Feb 2017, http://nla.gov.au/nla.news-article38883185

1932 'Black Cockatoos', *Sydney Mail (NSW : 1912 - 1938)*, 24 August, p. 16. , viewed 02 Feb 2017, http://nla.gov.au/nla.news-article166225182

1945 'AUSTRALIAN BLACK COCKATOOS BRED IN CAPTIVITY', *Townsville Daily Bulletin (Qld. : 1907 - 1954)*, 26 May, p. 5. , viewed 01 Feb 2017, http://nla.gov.au/nla.news-article61952758

1927 'WOOD-CHEWERS.', *Maryborough Chronicle, Wide Bay and Burnett Advertiser (Qld. : 1860 - 1947)*, 27 September, p. 8. , viewed 01 Feb 2017, http://nla.gov.au/nla.news-article150968558

1917 'NATURE NOTES AND QUERIES.', *The Argus (Melbourne, Vic. : 1848 - 1957)*, 5 January, p. 5. , viewed 01 Feb 2017, http://nla.gov.au/nla.news-article1589541

1948 'WILD NATURE'S WAYS', *Townsville Daily Bulletin (Qld. : 1907 - 1954)*, 18 September, p. 5. , viewed 01 Feb 2017, http://nla.gov.au/nla.news-article63367857

1912 'NOTES AND QUERIES.', *The Register (Adelaide, SA : 1901 - 1929)*, 16 March, p. 9. , viewed 01 Feb 2017, http://nla.gov.au/nla.news-article59053307

1890 'NATURAL HISTORY ITEMS.', *The Australasian (Melbourne, Vic. : 1864 - 1946)*, 26 July, p. 29. , viewed 01 Feb 2017, http://nla.gov.au/nla.news-article139140583

1933 'BUSH NOTES.', *The Australasian (Melbourne, Vic. : 1864 - 1946)*, 26 August, p. 44. (METROPOLITAN EDITION), viewed 01 Feb 2017, http://nla.gov.au/nla.news-article141379592

1933 'IN BUSH CORNERS', *The Sun (Sydney, NSW : 1910 - 1954)*, 30 May, p. 8. (FINAL EXTRA), viewed 01 Feb 2017, http://nla.gov.au/nla.news-article228889909

1938 'BLACK COCKATOOS, MYSTERY BIRDS', *The Sun (Sydney, NSW : 1910 - 1954)*, 27 September, p. 4. (LATE FINAL EXTRA), viewed 01 Feb 2017, http://nla.gov.au/nla.news-article230810497

1942 'NATURE NOTES.', *Townsville Daily Bulletin (Qld. : 1907 - 1954)*, 4 March, p. 2. , viewed 01 Feb 2017, http://nla.gov.au/nla.news-article63574227

1952 'OUR BLACK COCKATOOS', *Daily Mercury (Mackay, Qld. : 1906 - 1954)*, 15 March, p. 8. , viewed 01 Feb 2017, http://nla.gov.au/nla.news-article172091264

1928 'NATURE NOTES.', *Daily Mercury (Mackay, Qld. : 1906 - 1954)*, 14 December, p. 9. , viewed 01 Feb 2017, http://nla.gov.au/nla.news-article169953207

1928 'BLACK COCKATOOS.', *Daily Mercury (Mackay, Qld. : 1906 - 1954)*, 22 December, p. 5. (CHRISTMAS SUPPLEMENT), viewed 01 Feb 2017, http://nla.gov.au/nla.news-article169950653

1943 'BLACK COCKATOOS.', *Daily Mercury (Mackay, Qld. : 1906 - 1954)*, 6 November, p. 3. , viewed 01 Feb 2017, http://nla.gov.au/nla.news-article170988195

1934 'BLACK COCKATOOS', *The World's News (Sydney, NSW : 1901 - 1955)*, 31 October, p. 12. , viewed 01 Feb 2017, http://nla.gov.au/nla.news-article136999507

1924 'Black Cockatoos.', *The Land (Sydney, NSW : 1911 - 1954)*, 11 April, p. 22. , viewed 01 Feb 2017, http://nla.gov.au/nla.news-article103022630

1925 'Black Cockatoos.', *The Land (Sydney, NSW : 1911 - 1954)*, 24 July, p. 3. , viewed 01 Feb 2017, http://nla.gov.au/nla.news-article103037792

1913 'BLACK COCKATOOS.', *The Sydney Morning Herald (NSW : 1842 - 1954)*, 26 February, p. 9. , viewed 01 Feb 2017, http://nla.gov.au/nla.news-article15401163

1890 'BLACK COCKATOOS.', *The Australasian (Melbourne, Vic. : 1864 - 1946)*, 9 August, p. 25. , viewed 01 Feb 2017, http://nla.gov.au/nla.news-article139141169

1927 'BLACK COCKATOOS.', *North West Champion (Moree, NSW : 1915 - 1954)*, 10 October, p. 5. , viewed 01 Feb 2017, http://nla.gov.au/nla.news-article185525985

www.ingramcontent.com/pod-product-compliance
Lightning Source LLC
Chambersburg PA
CBHW041107180526
45172CB00001B/154